Benito MUSSOLINI

Biography®

Benito MUSSOLINI

Jeremy Roberts

A&E®

TF
CB

Twenty-First Century Books
Minneapolis

For my grandfather, who saw it coming

Twenty-First Century Books
A division of Lerner Publishing Group
241 First Avenue North
Minneapolis, MN 55401 U.S.A.

Website address: www.lernerbooks.com

Library of Congress Cataloging-in-Publication Data

Roberts, Jeremy, 1956–
    Benito Mussolini / by Jeremy Roberts.
        p.   cm. — (A&E biography)
    Includes bibliographical references and index.
    ISBN-13: 978-0-8225-2648-3 (lib. bdg. : alk. paper)
    ISBN-10: 0-8225-2648-4 (lib. bdg. : alk. paper)
    1. Mussolini, Benito, 1883–1945—Juvenile literature. 2. Heads of state—Italy—Biography—Juvenile literature. 3. Italy—Politics and government—1914–1945—Juvenile literature. 4. Fascism—Italy—History—20th century—Juvenile literature. I. Title. II. A&E biography (Twenty-First Century Books (Firm))
    DG575.M8R545 2006
    945.091'092—dc22   2005004219

Manufactured in the United States of America
1 2 3 4 5 6 – BP – 11 10 09 08 07 06

CONTENTS

Dictator Benito Mussolini wearing military gear about 1935. He ruled Italy from 1922 to 1943.

INTRODUCTION

The ten-year-old boy was not well liked at the boarding school. The teachers seemed to have it in for him. His classmates picked on him because he was poor. Every day someone reminded him he wasn't as good as they were. One day, when yet another student began insulting him, the boy couldn't stand it any more. He wasn't going to be a helpless victim. He pulled a knife from his pocket and stabbed the other boy in the hand.

"Your soul is black as soot!" yelled a teacher, grabbing him. "Tonight you will sleep outside with the dogs!"

And so, young Benito Mussolini went out into the cold night. Eventually, another teacher brought him inside. But the experience was one he never forgot. To Benito, it proved he must stand up for himself—and that doing so would mean fighting against great odds. Even when he was right, he believed, others would say he was wrong. As he grew older, he came to think that fighting and violence were justified if someone wanted to succeed.

Mussolini would use these lessons for the rest of his life. His willingness to fight for what he wanted helped him achieve many things. In time, it made him Italy's dictator and one of the most powerful men in the world.

But these lessons also led him to hurt many innocent people. As his power grew, so did the number of people harmed by it. Until, in the end, his soul truly was black as soot and he was cursed by much of the world.

The first Italian parliament meets in 1861, representing a united Kingdom of Italy.

Chapter **ONE**

THE MAKING
OF A DICTATOR

BENITO MUSSOLINI WAS BORN ON JULY 29, 1883, in a tiny northern Italian village called Varano di Costa and located in the district of Predappio. Like most of nineteenth-century Italy, the region was a rural area where agriculture was the most important industry.

Italy was still a young country at the time. For centuries the Italian peninsula—a boot-shaped piece of land jutting into the Mediterranean Sea—had been divided into many small kingdoms and city-states. City-states were tiny, independent countries centered on cities such as Venice. Other areas of the peninsula were ruled by foreign kings and the Roman Catholic pope. No single leader controlled the region. But during the mid-1800s, the Italian city-states united. They kicked out foreign rulers and took over the pope's territories, and in 1861 the unified Kingdom of Italy was born.

At the same time, a period known as the Industrial Revolution brought other changes. Large factories opened in cities such as Bologna, not far from Varano di Costa. Many Italian peasants (farmworkers)—who usually worked for low wages on land owned by the rich— left their difficult labor in the fields and took jobs in the factories instead. Meanwhile, many people, including Benito's parents, began fighting for more rights.

THE MUSSOLINI FAMILY

Benito's father, Alessandro Mussolini, was a blacksmith. His mother, Rosa Maltoni, Mussolini was a schoolteacher. Though not rich, they were considered respectable community members. Alessandro was very active in local politics. He was a Socialist and often expressed his controversial views. In general, Italian Socialists like Alessandro believed that the government should control the manufacture of goods, usually by owning the factories where goods are made. They also wanted better working conditions and believed that the workers—not rich factory owners—should be in charge of society. Alessandro named his son Benito Amilcare Andrea Mussolini after three Socialist heroes: Benito Juárez, a Mexican revolutionary who had fought Spanish control in the mid-1800s; Amilcare Cipriani, who had helped unite Rome with the rest of Italy; and Andrea Costa, who organized Socialist strikes and riots in the late 1800s.

Alessandro Mussolini, Benito's father, was active in politics. Alessandro believed strongly in Socialism, and his ideas influenced Benito.

Alessandro served in the local government for many years. But his politics sometimes got him in trouble. He had a bad temper and often became angry with people who disagreed with him. In 1878 the police warned him not to continue threatening to damage the property of people he disagreed with. And though he did not take part in it, Alessandro was arrested and jailed after Socialists began a riot in 1902.

Benito's mother taught elementary school in the village where the Mussolinis lived. Like many rural Italian schools, hers had only one room where students of all grades would be taught together. Conditions were difficult. When a window broke, Rosa had to beg the town council for money to fix it. Her job didn't

SOCIALISM AND COMMUNISM

The Industrial Revolution dramatically changed western Europe in the 1800s and early 1900s. While there were many benefits, some of the worst effects of the revolution were the concentration of poverty in cities and the dismal factory conditions. Among the many ideas to solve these problems was Socialism. According to this political philosophy, if workers owned the factories and controlled the government, they would no longer be poor or mistreated.

Socialism included a wide range of ideas and different specific plans. One of the most influential came from Karl Marx and Friedrich Engels, who published *The Communist Manifesto* in 1848. This book called for a violent revolution by workers, who would take over the country. The government they formed would then take over all industry and agriculture in the country. It would provide services such as housing, schooling, and health care for all.

Marx's theories, labeled Communism, influenced Socialism, but many Socialists believed they were too extreme. The call for violent revolution, especially, troubled many. Communism was used as a model by revolutionaries in Russia in 1917. After the Communists took over Russia's government, Communism gained popular support in other places in Europe, including Italy. Communists would eventually split from the Italian Socialist Party and form a separate group in 1921.

This painting of Benito's mother, Rosa Maltoni Mussolini, was created in the early 1900s. Rosa was very religious and often took Benito to church with her. But Benito did not share his mother's love of religion.

pay much but included two rooms near the school for the family to occupy.

Rosa, a Roman Catholic, was very religious. She often took her son to church and to religious festivals. Benito hated the smell of incense and the loud "drone" of the church organ. But he loved his mother deeply and usually did as she asked.

A second son, Arnaldo, was born to the Mussolini family in 1885. A daughter, Edvige, was born in 1888. Benito and Arnaldo shared a bed in the small, two-room apartment. Only a year and a half apart in age, the boys were very close.

SCHOOL AND WORK

Both of Benito Mussolini's parents valued education. When he was nine, they sent him to a boarding school in a nearby town. Students lived there full-time during school. The Catholic priests who ran the school enforced very strict rules.

Benito did not do well. "I was not a good boy," he admitted later. Though intelligent, he was a restless, unmotivated student. He tended to work hard only on subjects that interested him, and he often got into fights. He felt that the teachers picked on him. Finally, when he was eleven, he was asked to leave.

This time his parents sent him to a state school in Forlimpopoli, another small town in the region. The teachers here were not priests, and Benito's classmates included other teachers' children like himself. While he still occasionally got into fistfights, he did very well

Benito during his teen years. After a poor start in a Catholic grammar school, his parents enrolled Mussolini in a nonreligious school. He did much better there, discovering a talent for writing and speaking.

in his classes and was able to pass the exams required to get into the Collegio Giosuè Carducci, a school in Forlimpopoli. He studied math, agricultural science, and other subjects, but his favorites were the arts. He did well in literature and loved music. He also discovered a talent for writing and giving speeches.

Benito graduated from Giosuè Carducci in 1901 with a diploma that allowed him to teach elementary school. Like many graduates, Benito found himself hunting for work. In February 1902, he took a schoolmaster's job in a small elementary school—similar to his mother's—about seventy-five miles from home. But the new schoolmaster had problems from the start. Benito believed that the textbooks were terrible and made it hard for him to inspire his students. Some townspeople thought he was too easygoing with his students. Others said he spent too much of his free time playing cards and drinking. But perhaps his most serious trouble came when he started dating a married woman. It was a rocky and sometimes violent love affair. During one argument, Benito stabbed her in the arm with a knife. When school ended in June, he was told he would not be hired again.

ESCAPE

At the start of summer in 1902, Benito was eighteen years old. He was smart, decently educated for the time, and spoke well. He was interested in politics and poetry, and enjoyed having a good time. He loved

music and could play the violin well enough to enter-
tain himself and friends. He even had dreams of writing music. But Benito was also restless and
dissatisfied. He wasn't sure what he wanted to do with
his future. There seemed little opportunity for young
men like him in rural Italy.

Many other people felt the same way. Between 1896
and 1914, more than ten million Italians emigrated
(moved to other countries). A great number went to
the United States, seeking greater opportunities and
better lives.

Benito chose to move northward to Switzerland. He
would be able to further his education there by
attending lectures on different topics. He may also
have wanted to avoid the draft, as Italian law required
him to join the army at the age of nineteen. But most
of all, he wanted adventure.

Before Benito left, his mother gave him some money
to live on, but the small amount was soon gone. He
tried doing hard, physical labor for a very short
time—and didn't like it one bit. He took other jobs,
even begging and sleeping on park benches when he
ran out of money.

Eventually, he found that he could earn money writing articles for a Socialist newspaper. But because of
his poverty and because he expressed Socialist views,
he often got into trouble with the law. He was
arrested, jailed, and kicked out of one town after
another. Nevertheless, writing gradually became his

best source of income. It was easy for him and paid much better than hauling bricks or digging ditches. Most of his articles were very political. Usually he criticized existing society and called for change. He wrote that kings should be overthrown. He was also against the Catholic Church, believing that the priests and bishops who led it stood in the way of justice and reform. His writing got better and sharper as the months went by. "I can only write fiercely and resolutely," he later said.

Although he lived in Switzerland, Benito was still very much an Italian. He constantly thought about Italy. Although he had not been forced out, he considered himself an exile, since he would be arrested for draft dodging if he did return. Then, in 1904, Italy's king, Victor Emmanuel III, offered a pardon stating that draft dodgers who came home and agreed to join the army would not be jailed. For Benito, the pardon was a good reason to go home. His future, he was certain, lay in the country where he had been born.

A Son's Grief, a Soldier's Life

Benito Mussolini knelt at his mother's bedside in February 1905. He had rushed home from the army when he heard she was dying. Rosa had caught meningitis, an infectious disease that had no cure at the time. She lay in her bed, wasting away, unable to speak.

Mussolini apologized to his dying mother for the wild ways of his youth. He vowed to reform and asked

This photograph of Mussolini was taken early in his military service. Shortly after Mussolini joined the military, his mother Rosa died.

for her blessing. His mother squeezed his fancy new army hat with its feathery plume and held it close to her. Within a few days, she was dead.

"The only living being whom I had really loved and who had been near to me had been torn from me," wrote the twenty-one-year-old Mussolini. He mourned her death for many months.

But Mussolini still had a responsibility to serve as a soldier. Although he had originally been reluctant to join the army, Mussolini found he loved the discipline of military life. He belonged to an elite or special group of soldiers called *bersagliere*. Like modern U.S. Delta Force soldiers, these troops were trained to fight in small units and dangerous situations. The bersagliere had helped liberate Italy from foreign rule in the 1800s. "I liked the life of a soldier," said Mussolini, claiming that following orders and accepting discipline "suited my temperament."

Mussolini served with the army for about two years. In 1906 he went back to teaching. He did much better as a teacher this time and was popular with his students. But politics remained an important part of his life and continued to cause friction with local authorities. His restlessness led him to Trent on the Italian-Austrian border in 1909, where he got a job as the editor of a local Socialist newspaper. The city was ruled by Austria at the time, and Austria itself was part of the Austro-Hungarian Empire, a large collection of states and countries ruled by one emperor and bordering Italy in the northeast. But Trent had very close ties to Italy, and many of its residents considered themselves Italian.

THE "BIG" CITY SOCIALIST

By modern standards, Trent was a relatively small town, with barely thirty thousand people. But it was a large city to Mussolini, offering cafés, nightclubs, and other attractions. The twenty-five-year-old quickly came to love it. He wrote constantly and gave many speeches about the value of Socialism. He also spoke and wrote about the importance of being Italian and what he called "Latin genius and courage."

"Latin" referred to the language and culture of ancient Rome, which had dominated Italy centuries before. Mussolini, like many Italians, viewed the Roman Empire as a model of greatness.

Most of the people in Trent's government did not like Socialists. They saw Mussolini as a troublemaker. In September of 1909, he was arrested for theft, a charge that was probably invented to make it easier to put him in jail. He was found not guilty, but he was exiled from the city for his Socialist activities. He soon found another job as a newspaper editor in Forlì, a small northern Italian town near where he had grown up.

RACHELE

Soon after Mussolini arrived in Forlì, he began dating a young woman named Rachele Guidi. He had first seen Rachele nearly ten years before, while he was substitute teaching for his mother at her school. Although Rachele was seven years younger than Mussolini, her pretty blue eyes and blond hair caught his eye. A relationship soon blossomed between them.

Rachele's mother, Anna, tried to keep them apart. One night in the Guidi family kitchen, Anna told Mussolini, "Rachele is still under age. If you won't leave her alone, I shall lodge an official complaint against you, and you'll be sent to jail."

Mussolini left the house, only to return a few minutes later with a pistol.

"Now it's my turn to warn you," he shouted. "You see this revolver, [Mrs.] Guidi? It holds six bullets. If Rachele turns me down, there will be one bullet for her and five for me."

"Within two minutes it was all decided," recalled Rachele years later. "I was engaged to Benito."

In 1910 they moved in together in Forlì. A short time later, Rachele discovered that she was pregnant. The couple did not get married, however. Mussolini and Rachele believed that Socialist philosophy forbade formal marriage by the church or government. Committed couples were supposed to live together as husband and wife without needing a ceremony.

Rachele's family was poorer than Mussolini's, and she wasn't as well educated as he was. She knew little about the world beyond her tiny village. When they moved in together, she had never seen an umbrella. She was, however, extremely patient with his moods. And she remained faithful to him, despite the fact that Mussolini, who had always dated many women, continued to have affairs.

The couple's first daughter, Edda, was born September 1, 1910. When she cried at night, Mussolini sometimes took out his violin and played songs to soothe her to sleep. But most of his time was spent not with his new family, but working at the newspaper, giving speeches, and encouraging people to become Socialists. Handsome and energetic, Benito Mussolini was a familiar face in the town's central coffee shop.

Every night at eleven o'clock, he sat down at a table there to work on articles for his newspaper. His local fame and importance grew.

In 1911 the Italian government declared war on Turkey. Europe and the Ottoman Empire (centered in Turkey) had fought each other for centuries for control of the Mediterranean Sea. But the Ottoman Empire had grown weak, and the Europeans saw a chance to take over areas that had been under Turkish control. Italy especially wanted the North African territory of Libya and went to war to win it.

Socialists around the country opposed the war as unjust. Mussolini called it "mock-heroic madness of the war-mongers." He urged others to join a general strike (one that calls for the stoppage of all work) in late September to protest the war. Whipped up by Mussolini and others, local Socialists fought off Italian mounted troops sent to stop demonstrations. They destroyed train lines used by troops, overturned a streetcar, and ripped down telephone wires. Eventually, the Italian government made peace with Turkey—in part because of the unrest the war had caused at home.

ARREST

On the night of October 14, 1911, Benito Mussolini had many reasons to feel satisfied with himself. The Socialist Party was finally influencing the Italian government. He had a young daughter and a devoted wife. While not rich, he was an important man in town.

Mussolini in the early 1900s. By 1911 he had become an important man in Socialist politics. Many people knew and respected him, calling him professore *(professor).*

Strangers and others usually called him professor as a sign of respect for his learning. No one could have blamed him if he spent the rest of his life here as a politician and editor.

That night after he sat down at the café as he always did and began writing, two policemen came to his table.

"Professor Mussolini?" said one. "You must accompany us to the station."

Mussolini asked the officers if he could finish his coffee. They said of course. And then they marched Benito Mussolini off to jail. He was under arrest for supporting the strike and urging rebellion against the central government.

Mussolini about the time he took a job with the Socialist newspaper Avanti!

Chapter **TWO**

FASCISM'S BEGINNINGS

ONLOOKERS IN THE COURTROOM LISTENED AS THE prisoner finally got a chance to speak for himself. "If you condemn me you will do me honor, because you are not before a vulgar criminal," said Benito Mussolini. "[I am] an agitator and soldier of a faith."

Mussolini's faith was Socialism. He believed that it was the voice of the people and should be respected. But Mussolini never stood a chance to win his case. He was found guilty of creating disorder and encouraging riots. He was sentenced to a year in jail, but an appeal later shortened his term to five months. Nevertheless, Mussolini's dramatic trial and thundering speech attracted the attention of Socialists throughout Italy.

When Mussolini was released in March 1912, he was offered an important job writing for the Italian Socialist Party's newspaper *Avanti!* (Charge Forward), and soon became the paper's editor and boss. He moved his family to Milan, the large industrial city in northern Italy where *Avanti!* was published. Along with Rachele and Edda, Rachele's mother Anna lived with them and helped take care of her grandchild, a common practice at the time.

Mussolini's fame grew rapidly in Milan. He greatly increased the newspaper's circulation and popularity and did a good job managing it, though it never earned much money. Some biographers call Mussolini "one of the greatest journalists of the time" because of the power of his writing.

According to Rachele, it was around this time that Mussolini first got the nickname il Duce (meaning "the duke," and pronounced "eel DOO-chay"). Many people, including Rachele, still called him professor. As the head of a newspaper, he was also entitled to be called *direttore* (director). But Socialists began calling him il Duce because he seemed to have the makings of an important leader. The name stuck.

Mussolini's speeches and articles were always in favor of Socialism. He preached revolution, often by violent means. Workers should throw off the chains imposed by factory and farm owners, he said. When the revolt was complete, the world would be just and workers would rule.

When he made a speech, Mussolini seemed to catch fire. He often grew emotional, shouting and even ranting. But while this style may have seemed natural and spontaneous to observers, it was actually carefully planned. Mussolini knew just what effect the tone and volume of his voice had. He also studied body language, or the way he posed his body as he spoke.

In 1913 Mussolini decided to run for the national Chamber of Deputies, Italy's congress. He lost badly. Still, his future seemed bright. Many Italians were looking for political change. They thought that the government's leaders were corrupt and favored the rich over the poor. Not all supported Socialism, but many were willing to listen to new ideas.

MUSSOLINI'S BREAK

In 1914 tension in Europe—which had been brewing for a long time—was at a fever pitch. It erupted in June, when Serbs in Bosnia assassinated the heir to the throne of the Austro-Hungarian Empire. This empire included Bosnia, but many Serbs living there wanted to break away and join nearby Serbia. Blaming Serbia for the murder, Austria-Hungary declared war. Russia announced that it would come to Serbia's aid. Germany then countered with a promise to side with Austria-Hungary, while France prepared to help Russia. Within days the major powers of Europe had drawn up sides. Germany invaded Belgium and France in early August, and Great Britain and Russia came to France's aid.

As World War I (1914–1918) erupted, Italy was caught in the middle. It had treaties with both Germany and the Austro-Hungarian Empire. But it was also friends with France and Great Britain. At first, Italy attempted to stay neutral. Both sides pressured Italy to join them. The government debated what to do.

Mussolini first favored neutrality. He wrote in favor of it during the early weeks and months of the war. Many Socialists believed that countries should not fight each other, since most of the people who got hurt were the poor and working classes. But by October of 1914, Mussolini began to change his mind. He wrote an editorial in *Avanti!* saying that neutrality was "backward-looking and immobilizing." Italy should join Socialists in other countries, he said, and go to war against Austria-Hungary and Germany.

Mussolini's article started an uproar, and the Socialist Party's executive board met to discuss the matter. Furious with their lack of support, Mussolini quit the party and his job as editor.

To this day, historians debate why Mussolini changed his mind about the war. The most influential Italian biographer of Mussolini, Renzo DeFelice, argued that Mussolini realized that Socialism was going nowhere. According to DeFelice, Mussolini realized that nationalism (the belief that one's nation is more important than others) was more powerful and popular with Italians. He may have broken with the party, but he was staying close to the people. Other experts counter that

many Italians were in favor of neutrality and that
Mussolini's decision came from within. Some also
point to his military service and his attraction to violence as possible reasons for his change.

Once out of the Socialist Party, Mussolini started a
newspaper. The paper, called *Il Popolo d'Italia* (The
People of Italy), was backed largely by rich business
owners who favored the war and opposed Socialism.
Their support made Mussolini's new job a curious
one. As a Socialist, he had always criticized the rich.
But now he was working with them and for them. The
people who gave money to run the newspaper
included people in the Italian government, which
Mussolini had opposed.

In December 1914, Mussolini wrote an article that
said people should form *fasci d'azione rivoluzionaria*
(groups or bands of revolutionaries). Over the next
few months, he continued to develop this idea. He
said that he was not calling for the start of a new
political party. Instead, he saw the fasci, or "bands,"
as groups of leaders who would teach workers about
social justice and the need for revolution.

WOUNDED

In May 1915, Italy joined the Allies (the forces led by
France, Great Britain, and Russia). Mussolini soon
rejoined the army and was sent north to the mountains near Austria, where Italy's fiercest battles took
place. Mussolini, who became a corporal during the

war, was assigned to a unit in the area of Isonzo. The mountains were six thousand feet high, and fighting sometimes took place in areas with slopes that were nearly straight up. Soldiers had to be roped together to move safely. "What we suffered the first months!" Mussolini later said in his autobiography (written with his brother Arnaldo's help). "Cold, rain, mud, hunger! . . . War, with its heavy toll . . . and with its terrific hardships, surprised us," he added.

During a leave from the army, Benito and Rachele were married on December 17, 1915. According to Rachele, they got married at least partly because another woman, Ida Dalser, was claiming to be his wife. Mussolini had once had an affair with Dalser, who bore him a son. But after admitting to the affair, Mussolini agreed to marry Rachele. The couple's first son, Vittorio, was born in September 1916.

In the winter of 1917, Mussolini was badly wounded during a training exercise. He was sent to the hospital, where he recovered enough to go home. But his fighting days were through.

"I was a good soldier," Mussolini remembered later. "I showed fortitude and energy. Only when possessed of those qualities can a man stand gunfire."

Mussolini returned from the army in June 1917. He went back to work and cared for his family. His third child, Bruno, was born in 1918. Mussolini brought new energy to *Il Popolo*, which had fallen on tough times during his absence. He continued to work on

This photograph of Mussolini was taken during World War I, shortly before the young soldier was wounded during a training exercise. The injury ended Mussolini's career in the army.

his ideas about the fasci and wrote many articles supporting the war and Italian soldiers.

World War I finally ended on November 11, 1918, with an Allied victory. Much of Europe had been destroyed by the war. Roughly 680,000 Italian soldiers and sailors had died, with nearly 1 million other Italians wounded. Biographer Laura Fermi wrote about how greatly the war changed many men. "Mussolini, like millions of other soldiers, got used to the sight of death," she wrote. Fermi and others believe that Mussolini and these other men no longer valued life as much as they had before the war. But Mussolini himself believed the war had proven that Italians had a great, "warlike tradition" and that Italy was a great nation, capable of great things.

FASCI AND BLACK SHIRTS

Citizens across Italy celebrated wildly when the war came to an end. On the day before the final cease-fire was supposed to begin, Mussolini gave a speech in Milan. As he finished, he saw some soldiers in uniforms listening from a nearby truck. They were *arditi*, special Italian soldiers who were easily recognized by the black pullover shirts they wore. These men were trained to use grenades and knives to attack enemy soldiers in trenches in the first waves of attacks. Their task was difficult and extremely hazardous. Arditi had a reputation for being brave and reckless. As Mussolini put it, "They threw themselves into battle . . . with a supreme contempt for death." The arditi—sometimes called Black Shirts— also tended to cause trouble in bars and cafés when they were not on duty.

Mussolini had defended the arditi in his newspaper. After his speech, he jumped into the truck with the soldiers, sharing some champagne. He spent the rest of the day drinking with them. The next morning, the soldiers presented him with their unit flag: a white skull and crossbones on a black background. He hung the flag behind his desk.

Meanwhile, Mussolini's ideas about his fasci continued to develop. After the war, a number of groups

FASCISM: THE NATION ABOVE THE INDIVIDUAL

 lthough Fascism began as a small movement led by one man, it soon grew powerful. But even as it spread to other countries, it was never clearly defined and meant different things to different people. Even Mussolini sometimes contradicted himself when explaining or following its policies. In general, however, Fascists believed that the well-being of the country or nation was more important than that of its individual citizens. Because of this, Fascists did not value personal freedoms such as freedom of speech.

When Fascist leaders gained power, their political party and the government essentially became one and the same. While some non-Fascists might be allowed to participate in the government, they were never given a chance to gain real power. In addition, Fascist leaders generally dealt harshly with opposition, often arresting people or sending them into exile. Fascists used secret police and violence to scare and silence critics.

Other Fascist policies included a stance against both Socialism and Communism. Under Fascist leaders, wealthy businesspeople continued to own property and factories—although the government played an important role in guiding the economy. Fascists also believed in a strong military and were in favor of wars to gain territory.

were formed around the country, usually by ex-soldiers. One, called the Fasci di Combattimento (veterans' bands), met on March 23, 1919, in Milan with Mussolini's help. A variety of people came to the meeting.

They included arditi, other soldiers, former Socialists, and businesspeople. Fifty-four joined the organization.

Mussolini viewed the fasci as the start of a whole new idea called Fascism. Fascism "was a movement . . . not a party," he said. He thought the *fascisti,* or Fascists—those who believed in Fascism—were beyond the old ideas of political parties.

Fascism was only one of many ideas about politics and government in Italy after the war. But Mussolini gradually won others to his movement through his speeches and newspaper. He appealed to different groups in different ways. To the arditi, he sang the soldiers' anthem, "Giovinezza" (Youth), at rallies. He pleased military veterans by denouncing Socialists for refusing to back Italy in the war. He called for law and order and the end of devastating strikes, an idea businesspeople especially supported.

In 1919 Mussolini ran for the Chamber of Deputies again as a representative from Milan. He argued for better working conditions, such as an eight-hour workday; for higher taxes on businesspeople who had profited from the war; and for laws allowing more men and women to vote. But Mussolini's arguments were not enough. He and the Fascists lost the election badly.

Socialist Victories

Socialists, unlike Fascists, did well in elections throughout Italy, taking 156 of the 508 seats in the Chamber of Deputies. While the party did not control

the government, it could influence important deci-
sions. But these victories made others fearful. Factory
owners and the rich, especially, worried that the
Socialists might one day take over the country just as
the Communists in Russia had, seizing factories and
private property. So the Socialists' strong showing in
the elections actually weakened them in the long run.

Not all of Socialism's opponents were wealthy. Many
workers didn't like the party's policies against the war.
Socialist strikes disrupted everyday life and made even
going from one place to another difficult for everyone.
Socialist attacks on religion alarmed many Catholics,
who made up the vast majority of the population. And
Socialist ideas about property and marriage chal-
lenged some of the long-held values of Italian life.

This opposition to the Socialists—as well as to Communists, who were growing more active in Italy—
helped Mussolini and the Fascists. Since he had
become a vocal opponent of Socialism, others looked
to him as a natural leader in the fight against them.
His followers, especially the arditi, aggressively con-
fronted Socialists. There were many fights, as both
sides used violence and intimidation against each
other. Fascists attacked and burned the offices of
Avanti! Socialists fired at Fascist funeral processions.

"The country was in desolation," remembered Mussolini. "Italy was in the claws of disorder and violence." While he was partly responsible for the chaos
himself, many Italians agreed with his opinion.

Mussolini, right center, *and his Fascist Black Shirts about 1919. Mussolini believed that Fascism should become Italy's dominant political force by any means—including violence.*

MORE VIOLENCE

Mussolini's own battles with the Socialists were mostly with spoken and written words. But he was also willing to think about using other means in his fight. The police arrested him in late 1919 for having a large supply of illegal guns and weapons. Charges were later dropped. When local Fascist groups used violence to intimidate Socialists, Mussolini not only supported them, he supplied them with weapons, including revolvers and grenades.

Meanwhile, the Italian economy was doing very poorly. Many people were out of work. The government was spending much more money than it took in, further devastating the economy.

As the political and economic chaos grew, Socialists took over factories in Turin and Milan. Prime Minister Giovanni Giolitti managed to control the crisis with patient, peaceful talk, and the Socialists left the factories. But there were other disruptions and clashes. Attacks on Socialists by Fascist groups grew. In Bologna Socialists won local elections and took control of city government in November 1919. The following year, a Fascist group led by a friend of Mussolini attacked City Hall. "The reality is this," wrote Mussolini. "The Socialist Party is a Russian army encamped in Italy. Against this foreign army, Fascists have launched a guerilla war, and they will conduct it with exceptional seriousness."

DANGER AND FAMILY

Despite his busy schedule, Mussolini took time out in 1920 to learn how to fly airplanes. The airplane was still a new invention, and he loved the thrill and adventure of flying. He would arrive at the airport dressed in a black business suit and round bowler hat, then climb aboard with his instructor. Sometimes he would bring Rachele and his sons to watch. In March 1921, Mussolini crashed, but he escaped with only a few cuts. He had also developed a taste for very fast cars and loved to drive at high speed on the winding local roads.

In addition to his love of danger, Mussolini had not lost his terrible temper and his willingness to fight.

He often engaged in duels. These sword fights were generally guided by very strict rules—which were supposed to guarantee that they did not end in death. However, injuries were common, and Mussolini's daughter Edda remembered that her father often returned home from duels with wounds. But he also always had something for the children—including, once, a stray kitten that he thought had brought him luck.

In general, Mussolini was an easygoing parent who indulged his children. "My mother... was the real dictator of the family," remembered Edda. Rachele Mussolini was a strict mother with all the children. If they misbehaved, they were spanked—or, as Edda put it, "soundly thrashed!"

CHANGES

In the spring of 1921, the country once more held elections. This time, Mussolini's Fascist movement—though still not an official party—managed to get thirty five followers elected to the Chamber of Deputies, including Mussolini. It was a small group but an important win for the Fascists.

However, as the movement grew, Mussolini began to have trouble controlling the various groups within it. There were many reasons for this difficulty. Some local leaders were jealous of his influence and wanted power for themselves. Others had disagreements over policy or what the Fascists should do. One of the most important disagreements came during the summer of

1921, when Mussolini suggested that the Fascists make peace and cooperate with the Socialists.

Historians continue to develop theories about Mussolini's thinking on this issue. Perhaps he recognized that the Fascists could not continue to use violence if they wanted popular support. Perhaps he thought the movement would go further if it worked with the Socialists on some issues. Perhaps he was just pretending to be cooperative to gain backing from others who did not like violence. In any event, most Fascists disagreed with the stance, and they held a meeting in Bologna to criticize Mussolini. And there, rather than fighting as he had all his life, Mussolini did something completely unexpected: on August 18, 1921, he resigned from the executive committee of the Fasci di Combattimento.

If his resignation was designed to show how important he really was, it worked. Within days, other Fascists were urging him to return. Mussolini spent the next three months developing his movement. By November 1921, he had reorganized fascism into a more traditional political party. At a national meeting that month, Mussolini was elected to lead the new Fascist Party—although he did have to officially take back his calls to make peace with the Socialists. And by now, all of the Fascists called him il Duce.

THE MARCH BEGINS

By the summer of 1922, economic problems, bank failures, and the collapse of the government under

Prime Minister Ivanoe Bonomi led to a series of protests from Italian workers. In response, bands of well-organized Fascists marched through large areas of Italy, destroying the homes of Communists and Socialists as they went. The new government, made up of many political parties with different ideas of how to govern the country, was too weak to stop the violence. Some politicians who hated Socialism even backed the Fascists. The Black Shirts—still made up mostly of arditi but also including other Italians—fought openly with Socialists and anyone else who opposed them.

The Fascists had an opportunity to boost their public image in August, when the Socialist Party staged a general strike. The government—still too crippled by its problems to act—was unable to do anything to

Black Shirts burn Socialist literature in 1922. The Fascists would take destruction on the road during the March on Rome that year.

stop the devastating strike. Fascists stepped in and took over many of the most critical jobs, such as delivering mail and protecting power stations. The strike soon collapsed, and Fascists got the credit.

In October Mussolini's newspaper published regulations outlining the formation of a Fascist militia, a private army equipped with weapons. This move was clearly against the law, but again the weak government did nothing. The Fascists had three hundred thousand members, and the militia's violent ways and Mussolini's fiery speeches to noisy, excited crowds made the movement seem much larger. On October 24, 1922, the party held a meeting in Naples, a large city in southern Italy where fascism had not been very popular so far. The party found unexpected support there, and the group suddenly swelled. Farmers and rural workers in the south had been hurt by the bad economic times. They were angry with the government that seemed unable to help them. They were also angry at the Socialists, who—it seemed to them—had caused a lot of trouble and only made things worse. Within a few days, the Fascists began marching from Naples and other cities toward the capital city of Rome, demanding a Fascist government.

Mussolini later claimed to have planned and organized the march. While some scholars have questioned this, it is clear that he quickly took control of the situation. Within a few days, Italy was in the middle of a revolution. And perhaps the person who best understood what was truly happening was Benito Mussolini.

Mussolini, fourth from right, *joins members of the Fascist militia and other supporters during part of the March on Rome in 1922.*

Chapter **THREE**

IL DUCE

MUSSOLINI TOOK HIS SEAT IN THE MILAN THEATER, smiling and waving at acquaintances. He seemed completely unaware of the chaos in the rest of the country. Yet at that very moment, Fascists from different parts of Italy were marching toward Rome. Italian citizens were filled with questions: Were there going to be riots? Was the army going to attack the marchers? Would Italy collapse? And the biggest question of all—what would the leader of the Fascist movement do?

The answer, at least for tonight, was this: ignore the questions and turmoil and go to see a play.

Mussolini did love the theater, but that night he was as much of an actor as anyone on the stage. He was pretending to be calm, hoping to pressure the government in Rome. The Fascists' march on the capital was a daring gamble. It might make government leaders give Mussolini and his followers a bigger say in the government. On the other hand, if it failed or fell apart, no one would ever take Mussolini seriously again. Mussolini believed that by pretending not to care, he would seem as if he was in total control. In the meantime, the crowd of Fascists, estimated at about twenty thousand to twenty-five thousand, reached the outskirts of Rome.

Mussolini's instinct was right. On October 29, 1922, just as a hard rain began driving many Fascist marchers

Mussolini, fourth from left, *and other Fascist officials listen as Italy's King Victor Emmanuel III names them to the new Italian parliament in 1922. The king made Mussolini prime minister, giving him control of parliament.*

home, the king—hoping to restore order—asked Mussolini to become the prime minister. Mussolini later told a writer that he felt like an "artist" at the beginning of his career, anxious to begin a lifetime of creative works.

PRIME MINISTER

In the Italian system of government at the time, the king was technically the head of the country. However, his powers were very limited. The prime minister, on the other hand, had a great deal of say in what the government did. He headed a cabinet (group of advisers) made of different ministers who watched over different government functions, such as collecting taxes or dealing with foreign governments. And he had a great deal to say about the laws the Chamber of Deputies passed and how they were enforced.

Besides being prime minister, Mussolini also held the jobs of foreign minister and interior minister. As interior minister, he had direct control of the national police. He could also authorize wiretaps on phones to find out what his enemies were saying. And he had some secret funds that were not reviewed by the legislature and that he could use however he saw fit.

By holding all of these posts, Mussolini had more power than previous prime ministers. But his power was not absolute. Of the fourteen cabinet members Mussolini appointed, only four were members of the Fascist Party. There were several key reasons for this.

QUESTIONS FROM HISTORY

Fascists killed approximately two thousand people in the months before Mussolini became prime minister. During the Fascist rise to power, the group used illegal means such as assault, arson, and murder to take a legal place in the government. Party members, from Mussolini down, openly endorsed violence as a legitimate tactic, rather than condemning it as an illegal and undesirable act.

Historians have struggled to explain why violence was relatively acceptable and even attractive to both Fascists and non-Fascists in Italy at the time. Some think that World War I made people less sensitive to violence. Others suggest that bloody revolutions in other countries made violence seem more tolerable. Some say that, in such chaotic times, violence seemed the only way to bring order to society. It may be that the answer includes all of these reasons.

Another question is why the government didn't try to stop Mussolini. An army attack on the marchers, some say, could have ended the entire Fascist movement. On the other hand, others point out that Mussolini's movement appealed to many soldiers, and government leaders may have been afraid that the army would mutiny (rebel). The government leaders also may have thought they could control Mussolini, or get rid of him after a few months. After all, many government officials had come and gone in the short time since the war. They did not fully understand Mussolini's hunger for power or how willing he was to use violent means to keep that power.

Socialists did oppose Mussolini's appointment. But they were a minority in the Chamber of Deputies. And just before Mussolini became prime minister, their newspaper's offices were burned— preventing them from publishing calls for a general protest or strike. The fire was probably set by Fascists, at Mussolini's order.

One was practical: Mussolini needed to work with the other parties if he hoped to get any laws passed. Naming members of other parties to the cabinet helped guarantee support in the Chamber for laws that Mussolini wanted. The Fascists also had little experience in many government areas.

This approach to government was not new. Italian cabinets and governments had a long history of different factions working together. So, from the point of view of other politicians, Mussolini was simply continuing tradition and working legally—despite his speeches calling for a Fascist revolution. It is possible that Mussolini's fellow politicians did not see him as very different from them because he hid activities that others might question. But he also did use legal means to run the country.

Meanwhile, Mussolini steadily increased his power. One of the first things he set out to do was enlarge Fascist representation in the Chamber of Deputies. Using his support in the cabinet, he had a new law passed that promised a large majority of seats to the party that won the most votes in a general election. New elections were then scheduled for the Chamber of Deputies.

Mussolini also established a Fascist Grand Council, a kind of executive board to help him run the government. The council dissolved the Fascist Black Shirt squads, along with all illegal militia groups in the country. In their place, Mussolini established a voluntary militia for national security. This group was really just renamed Black Shirt squads—except now the squads were legal.

The Fascist Party could use them to harass enemies or to enforce any of Mussolini's orders.

CORFU

Prime Minister Mussolini soon faced his first major international issue. In August 1923, an Italian delegation trying to settle a border dispute between the nearby European countries of Greece and Albania was ambushed and killed. Mussolini—who had already had disagreements with the Greek government—blamed Greece for the attack. He presented the Greek government with a long list of demands, including the death penalty for the murderers, payment of a large fine, and demands that Greek officials honor the Italian flag in a ceremony in an Athens Roman Catholic Church. When Greece did not agree to all the items on the list within twenty-four hours, Mussolini acted swiftly. He sent seventeen warships with marines to occupy Corfu, a small island between Albania and Greece. More than a dozen people died when the ships opened fire on Corfu.

The matter was taken to the League of Nations, a world organization established after World War I to settle international disputes. Mussolini refused to accept the league's authority and threatened to pull Italy out of the league. But a separate group of ambassadors helped the two countries come to an agreement. In the end, Greece paid the fine and Italian ships left Corfu.

Mussolini's actions impressed many Italians. He had demanded that other nations respect Italy—a demand that made a lot of sense to them. They had seen Greece's refusal to offer such respect during the Corfu affair as an insult to the country's pride that had to be avenged. But other world leaders and people outside of Italy had a different view of the Corfu incident and of the Fascists in general. They considered Mussolini a bully. Some thought Italy was falling apart and that he wouldn't last as prime minister. Still others thought his methods were necessary.

Some foreigners even looked to Mussolini and his movement as a model. In Germany Adolf Hitler hoped that his small Nazi Party would make him "Germany's Mussolini." But with only a handful of followers, he had a long way to go.

LANDSLIDE VICTORIES

In Italy, the local Fascist bands had not stopped attacking their enemies. Some well-known critics of Mussolini and the party were killed, including Catholic priest Giovanni Minzoni. Not only did this silence these critics, it intimidated others. That silence meant that few people knew about the dark side of Fascism. At the same time, many Italians joined the Fascist Party to preserve their influence in government. The result was a landslide for the Fascists in the 1924 elections. Mussolini's party won big, with candidates getting about half of the votes in the north and more than 80 percent in the south.

Meanwhile, Mussolini was so busy he hadn't even bothered to find a proper house to live in. He ran the government from luxury hotels in Rome, while his family remained in Milan. Finally, he bought an estate called Villa Carpena, located just outside his old hometown. He moved his family there while he stayed and worked in Rome. Rachele and the children enjoyed the change in status that the move brought. Rachele was treated as a signora, or a lady of respect, an honor that usually went to wealthy women whose families had owned estates for generations.

Mussolini had many party and government advisers around him in Rome, but his brother Arnaldo was by far his closest adviser and friend. Arnaldo remained in Milan, where he had taken over his brother's newspaper, *Il Popolo d'Italia*. The two men spoke every night, discussing anything and everything. Benito trusted and loved his brother deeply. He also relied on him to boost his ego. He called him, "the one to whom I turn first when I require unstinted devotion."

A steady stream of visitors came to see Prime Minister Mussolini at his offices. Mussolini had different ways of impressing them. He might show off a famous statue on display or talk about a topic that his aides had told him the visitor was interested in.

As a journalist, Mussolini knew the importance of good publicity. He made sure that photos of him doing "manly" things such as swimming, skiing, and horseback riding constantly appeared in the press. He tried to project the image of a strong, healthy man in the photos and in news releases about his activities. Party and government public relations experts helped get this positive message out. Many of the stories about his personal life either greatly exaggerated his good points or were simply false, written to impress people. Negative aspects, on the other hand—such as his many affairs with women—were never publicized.

MURDER

One June 9, 1924, Socialist Giacomo Matteotti, a member of the Chamber of Deputies and an important critic of Mussolini, was kidnapped from Rome. Police quickly traced the car used in the kidnapping to a group of Fascists led by Amerigo Dumini. On June 12, Dumini was arrested, while police continued to search for Matteotti. Matteotti's body was eventually found in a shallow grave. He had been killed, although the official findings cast doubt on whether his death had been deliberate or if an accident had occurred during the kidnapping.

Dumini had many friends among high-ranking Fascists, including Mussolini himself. Critics accused Mussolini of ordering the kidnapping—and maybe murder. A number of members of the Chamber of Deputies walked out of the parliament, declaring the government illegal. They accused Mussolini of using his position as interior minister to set up a secret police force and silence critics.

Mussolini responded by giving up his interior minister post and appointing a well-respected former politician, Luigi Federzoni, to the job. He claimed that the crime had been committed by rebel Fascists, and he made a point of making sure that Matteotti's widow received payments from the state. Dumini and others involved in the disappearance were arrested and put on trial.

But the turmoil continued. A Fascist member of the Chamber of Deputies was assassinated in the fall. Fascists pushed Mussolini to take strong action against their enemies or else face a second Fascist revolution— this time, one that would remove him from office. The Socialists who had walked out of the Chamber of Deputies did not return. Instead, they talked of armed resistance and starting a new government.

On January 3, Mussolini strode into the Chamber of Deputies and delivered a speech saying that he had nothing to do with the kidnapping and had not set up a secret police. He did not have a secret death squad and would not have used it to murder an opponent. But then, citing attacks by Communists and Socialists, including attacks on trains and police stations, he said that the situation was getting out of control.

"Italy, gentlemen, wants peace, wants quiet, wants work, wants calm," he told the Chamber, which now contained only his supporters. "We will give it with love, if that be possible, or with strength, if that be necessary."

Mussolini had, essentially, declared himself dictator of the country. He immediately sent the Fascist militia

DID HE KNOW?

To this day, experts debate whether Mussolini knew in advance that Matteotti would be killed. Some aren't even sure that the men who took him actually intended to kill him. They point out that his body was not well hidden and that the kidnappers' car apparently ran out of gas. But Mussolini certainly hated Matteotti. He also could have made sure that Dumini and others were more seriously prosecuted and received stiff jail sentences. In fact, the investigation into the affair seems to have been designed more to cover up details than reveal them. A trial led to the conviction of the suspects, but they went to jail only for short terms. And after Dumini got out of prison, he received a lot of money from the Fascist Party—with Mussolini's approval.

Many people are also not convinced that Mussolini had a secret assassination squad. Even if he didn't, however, Black Shirts such as Dumini were clearly ready and willing to hurt or kill opponents. At the very least, Mussolini had created an atmosphere where Fascists believed they were above the law.

out to battle protesting Socialists and to restore order. He presented a new cabinet of ministers to the king, removing most members of other parties. He also named himself head of the army, navy, and air force. This move gave him direct control over the military.

AN ASSASSIN MISSES

As Mussolini tightened his control of the Italian government, he made more and more enemies. On September 11, 1926, a man named Gino Lucetti threw a bomb at Mussolini near his office. Mussolini escaped without injury. Then, on October 31, as il Duce traveled by car to Bologna, Anteo Zamboni tried to shoot him. When the gun went off, Mussolini grabbed first his neck and then his head.

"They did not get me!" he yelled to his driver, and the car sped away. At the same time, Fascists in the nearby crowd attacked Zamboni and killed him.

Mussolini sports a bandage after surviving one of several assassination attempts in 1926. The attempts on his life did not stop Mussolini from seeking greater power and control.

Mussolini survived at least four assassination attempts in the early years of his reign. But rather than fear future attacks, he used these incidents to increase his power. His followers passed strict laws banning opposition parties, unions, and other groups. From then on, anyone who wanted to participate in Italian politics or government had to be a Fascist. People convicted of political crimes could be put to death. Newspapers could not criticize Mussolini or his government. Border police prevented anyone from leaving the country without permission. Local elections were banned, and mayors were now appointed by the government and were Fascist officials. Mussolini named a new national police chief, Arturo Bocchini. Bocchini created a secret police agency called OVRA (Organizzazione di Vigilanza Repressione dell'Antifascismo, or Organization for Vigilance against Anti-Facism) to enforce political laws. They could arrest anyone who criticized Mussolini. Such critics—who included Fascists as well as others—were often sent away to tiny villages in southern Italy and the nearby islands. They lived there as exiles, cut off from family and friends.

Il Duce was now the absolute dictator of Italy. He controlled the Chamber of Deputies, the police, and the military. His word was law. Italy had become a *stato totalitario*—a totalitarian state, in which one party and one leader held virtually all power.

Dictator Benito Mussolini dressed in his military uniform, complete with medals, gives a public speech shortly after becoming the dictator of Italy.

Chapter **FOUR**

POWER AND AGGRESSION

IN **TOTALITARIAN ITALY, SCHOOLCHILDREN USED** textbooks approved by the Fascist government. Six-year-olds learned to write by copying out sentences such as, "Long live Italy! Long live the army! Long live our Chief Benito Mussolini." The Fascist salute—an extended raised arm—was a mandatory part of most ceremonies. And anyone who angered local Fascists could expect to be punished—often by bands of angry thugs.

"Italy . . . is a silent and shadowy world where men are afraid to be seen in the company of the truth," wrote a British journalist in mid-1920s. The reporter noted that many Socialists had gone into hiding and that no one dared to say anything against the government. At the same time, he found that Italians accepted and even welcomed Mussolini. "There is no serious revolt hatching. . . . There is not even a general indignation."

For many Italians, life under Mussolini was better than it had been during the chaos of earlier years. Disruptive strikes were a thing of the past. Factories operated more smoothly. Trains, critically important for even short trips in a country where most people did not own cars, ran with far fewer interruptions. The Fascist Party sponsored summer camps for children that let them spend a month at the mountains or the seashore.

Some freedom of expression was still allowed, as well. Newspapers were often permitted to criticize certain Fascist policies—so long as they praised Mussolini himself and did not voice favor for Socialism or Communism. Artists, college professors, and other intellectuals were usually left alone. In fact, many joined the Fascist Party. Even after a law was passed requiring professors to swear an oath of allegiance to the Fascist Party, only a handful of those who refused to do so were fired.

Mussolini was well suited to his role as dictator, but the stress of governing hurt his health. He developed a stomach ulcer that left him spitting blood and unable to work. Doctors suggested surgery, but he refused. He changed his diet and stayed in bed for a few weeks, which helped, but the ulcer continued to plague him.

THE CATHOLIC CHURCH

One of Mussolini's goals as Italy's leader was to resolve a long-running dispute with the church. Since the final days of the Roman Empire, the Roman Catholic Church had held a unique place in Italy. For centuries, the pope, who headed the church, had dominated Italian politics. Wanting to hold onto that power, the church fought the creation of the new, unified government throughout the 1800s. In turn, modern Italy's founders seized Rome and the Papal States, which had been ruled and in some cases owned by the Catholic Church.

Nevertheless, the Catholic Church remained very influential in Italy. The vast majority of the nation's people were Catholics. The church had large sums of money, and its officials tended to be well educated and knowledgeable on a number of issues.

Church leaders remained at odds with Italian governments throughout the late 1800s and early 1900s. They wanted the lands that had been seized from them to be returned or at least paid for. They resented government actions such as taking over the right to marry people, abolishing laws that mandated religious training, and stripping the church of some rights to discipline priests.

As a young Socialist—and even in his early days as a Fascist leader—Mussolini had often criticized Christianity, the Catholic Church, and the pope. But as Italy's dictator, Mussolini wanted to reach an agreement with the church over the disputed lands and other issues. He knew that such an agreement would please most Italians.

For three years, government representatives and church officials negotiated in secret. When the two sides came close to an agreement, Mussolini himself got involved, working on the proposals for long hours. Finally, on February 11, 1929, Mussolini and Cardinal Pietro Gasparri, acting for the pope, signed the agreement, which became known as the Lateran Pact because it was signed at the pope's Lateran Palace in Rome.

Under the agreement, the church accepted the fact that it would not regain control of the territory that had been seized when Italy was created. However, it was paid for the property. Vatican City—a church-owned area centered in Rome's Saint Peter's Basilica and less than one-fifth of a square mile in area—was officially placed under church law, not Italian law. The Catholic religion was recognized as Italy's state religion, and the church regained authority over rites such as marriage.

The agreement was enormously popular with Catholics both inside and outside Italy. Pope Pius XI praised Mussolini, and the official Catholic newspaper proclaimed, "Italy has been given back to God and God to Italy."

A NEW HOME

In November 1929, Mussolini at last brought his family to Rome to live with him. They moved into a large country villa, or estate, called Villa Torlonia. During his spare time, Mussolini would ride horses and bicycles on the property. He also played tennis and spent time with his two youngest children, Romano (born in 1927) and Anna Maria (born in 1929). The children had a variety of pets ranging from turtles to dogs.

Edda, his oldest daughter, was now nineteen years old, an age when most young Italian women married. Her father did not approve of some of the young men who tried to date her, however. When she fell in love with a Jewish man, Mussolini warned her that marriages between people of different religions generally failed. That romance, like others, soon died. But in 1930, Edda fell in love with twenty-seven-year-old Count Galeazzo Ciano, the son of Fascist admiral Count Costanzo Ciano. Admiral Ciano had worked with Mussolini since before he came to power. His son now worked as a diplomat. Mussolini heartily approved. Four thousand people were invited to the estate on April 23 for a huge reception. The next day, the couple married in a small church not far away.

Mussolini now seemed to be as powerful in his country as the ancient Roman emperors had been in their empire. But some things were beyond even his control. On December 21, 1931, his brother Arnaldo suffered a stroke. Mussolini rushed from Rome to his brother's bedside, but he arrived too late. Arnaldo was dead. Mussolini spent the long night of December 22 weeping and watching over his brother's body. He had lost his closest friend—and perhaps the only person in the world he truly trusted.

CELEBRATION AND DEPRESSION

In the fall of 1932, Mussolini and Italy celebrated the tenth anniversary of his rise to power. He mounted a horse and rode in a magnificent procession down the Via del Fori Imperiali, a new road he had ordered to be built through the center of Rome to remind Italians of their glorious past. The Forum, a large area of ancient Roman government buildings, temples, and monuments, lay alongside the road. The remains of these massive marble structures date back to the time of Julius Caesar, a great leader who had constructed the first Roman forum in 46 B.C. For a man like Mussolini, who linked the glory of the past with the potential of the future, there was no better view in Rome.

A new challenge arose during the early 1930s, when the world plunged deeper and deeper into an economic downturn known as the Great Depression (1929–1942). Although the Depression initially hit the United States hardest, countries everywhere felt its effects. In Italy nearly 1.3 million people were out of work by February 1933.

Mussolini launched a series of programs designed to jump-start the economy. He also announced the creation of new corporations that would be owned by the state and would be used to boost industries ranging from agriculture to the arts. The program received a lot of newspaper coverage but actually had little impact on the Italian economy or the Great Depression. Yet,

Mussolini's willingness to try new ideas earned the admiration of many.

Elsewhere, the Great Depression brought turmoil and much greater hardship. Nowhere was this more true than Germany, where in early 1933, Adolf Hitler became the head of state. In many ways, his Nazi Party was a darker, more violent version of Mussolini's Fascists. Within a few months, Hitler's shadow would spread beyond the borders of Germany. As it grew darker and darker, even Mussolini could not afford to ignore it.

MEETING WITH THE FÜHRER

The countryside outside of Venice stretched before Mussolini as he walked on the massive mansion's lawn on June 14, 1934. The ornate palace had once belonged to Napoleon Bonaparte, who had conquered most of Europe as the leader of France in the 1800s. It was a majestic setting for Mussolini's meeting with Adolf Hitler, the new führer (leader) of Germany. The beauty of Venice and the June weather would provide a pleasant backdrop to their discussions.

The scene may have looked perfect from the distance, but up close it was less than ideal. Mosquitoes swarmed over the mansion. Venice's canals, beautiful to look at, smelled of raw sewage. The discussions with Hitler were much the same. From a distance, it was a perfect meeting of two of the most important leaders in Europe. Up close, there were many disagreements.

Adolf Hitler, left, *walks alongside Mussolini during their 1934 meeting in Venice, Italy. The men met to discuss relations between their countries and their visions for Europe's future. As is evident on the leaders' faces, many disagreements separated the two powerful men.*

Like Mussolini, Hitler used violence to rise to power. He had long admired Mussolini and his tactics. But he had struggled for far longer. Many more people had been killed during his rise. He believed that Jews and other groups he saw as "inferior races" should be removed from Germany. And he wanted to greatly expand German territory—starting with Austria.

Italy shared a border with Austria, and Mussolini was not eager for Germany to take over that country. He was worried that Germany might become too powerful if it held Austria. In addition, Mussolini liked Austria's leader, Chancellor Engelbert Dollfuss. He made these views clear to Hitler at their meeting.

HITLER

dolf Hitler was born in Austria near the German border on April 20, 1889. As a young man, he wanted to be an artist and even managed to sell a few paintings before he joined the German army in World War I.

After the war, Hitler became active in politics. He combined his extreme hatred for Jews with a political philosophy called National Socialism, or Nazism. Nazism shared many things with Fascism, including an acceptance of violence and a hatred of Communism. The system of government Hitler wanted was a dictatorship controlled by a single political party, just as in Mussolini's Italy.

Hitler and his tiny party attempted to take over the German government with an armed coup in 1923. They failed miserably. But Hitler's trial brought him to national attention—much as Mussolini's had years earlier. When he was let out of jail, he rebuilt his party. Economic problems in Germany led to Hitler's selection in 1933 as leader of the Reichstag, or German parliament. Hitler used this important position to increase his power until he became dictator of the country.

As Hitler rose to power, he encouraged Germans to believe they were superior to all other people and should rule Europe and the rest of the world. He soon began pushing to expand German territory, and he was willing to wage outright war to accomplish that goal. Hitler also brought an extreme anti-Semitism (hatred for and prejudice against Jews) to his politics. He believed that Germans were part of a superior race and considered Jews and other groups inferior. He had followers persecute and kill Jews as well as others. By the mid-1930s, Hitler and the Nazis had begun to emerge as a terrifying new force in Europe.

But Hitler talked about Austria for hours, boring Mussolini. Mussolini also disagreed with Hitler's attitudes toward Jews and other people who lived along the Mediterranean—including Italians—calling Hitler's philosophy "one hundred percent racism. Against all and everybody."

Still, the meeting was important. The two European dictators realized that, while they disagreed on some things, they had much more in common with each other than with European countries such as Great Britain and France, which remained democracies.

MISTRESS

Mussolini was concerned about Italy's image in the world. The country had to be tough and aggressive. As its leader, he tried to project the same image to his country.

Now fifty years old, Mussolini continued to pride himself on his manly appearance. With his hair thinning, he shaved his head. He often wore military uniforms and hats, even a helmet, especially when he gave speeches and met foreign leaders. At state ceremonies, he would sometimes ride a towering white horse and wear a metal helmet with a large eagle, the symbol of ancient Rome. He'd grown plumper over the years but still exercised by swimming and horseback riding when he could.

While he had many advisers in government, he did not consider them good friends. And with the exception of Edda, his oldest daughter, he didn't seem to be very close to his children. They respected him but also impressed friends by mildly defying him or criticizing him. "Daddy hasn't managed to do anything he wanted to do," bragged the teenaged Vittorio to a classmate.

In fact, Mussolini sometimes complained that, as the nation's dictator, he had to remain isolated from nearly everyone. He also took some blame. "I cannot have any friends," he told an interviewer. "First of all, because of my temperament; secondly because of my view of human beings." He simply did not trust people.

But Benito Mussolini did become emotionally close to a young woman named Clara (usually called Claretta) Petacci. Claretta came from a well-off Roman family. Her father was the pope's doctor. She was twenty-one years old—less than half Mussolini's age— and married. But when she met Mussolini at a beach resort, the two became friends immediately. No one is certain how soon their relationship became romantic, but eventually Claretta had her own apartment in the building that housed Mussolini's office in Rome. In the past, Mussolini had changed mistresses often. But with Claretta, he had found a deeper, lasting relationship.

ETHIOPIA

In July 1934, Austrian Nazis who admired Hitler murdered the head of the country, Chancellor Dollfuss, in an attempt to take over the government. Mussolini was outraged. Not only did he believe that Hitler had encouraged the crime, but Dollfuss's wife and family were visiting Mussolini's own family at the time. The murder seemed like a personal insult to the Italian leader. He sent four army divisions to the border as a warning to Germany not to interfere in Austria.

The rest of Europe was suspicious and worried about Hitler's Germany. Many looked at Italy with the same suspicion. The two dictators had used similar methods to come to power and run their countries. In many ways, the other nations were right to worry. For despite Mussolini's defense of Austria, he was planning a foreign invasion of his own.

Starting in the 1500s, European countries had established colonies in Africa. After Italy became a nation, its leaders also wanted a colony. In the 1890s, Italy claimed lands in North Africa, taking over Eritrea in the northeast, as well as nearby islands. In 1895 Italian troops invaded Abyssinia (later known as Ethiopia), just south of Eritrea. They were defeated, and Italy was forced out of the country. But Italy still controlled the other areas it had claimed in North Africa.

In 1934 Mussolini began thinking again of expanding Italy's African territory. Another colony might help the mother country's economy by adding new resources. Defeating another nation would also make Italians feel more important—and make them praise Mussolini. And always looking to Italy's glorious past, Mussolini also dreamed of reclaiming African lands once held by the Roman Empire.

Italian soldiers, protected by their army's tanks, push through African foliage during the invasion of Ethiopia in 1935. Mussolini had set his sights on expanding Italian territory in North Africa.

During the early months of 1934, Mussolini discussed the possibility that Italy would invade Ethiopia with the British and French governments, who had African colonies of their own. The French did not raise any objections, but the British made it clear that he should not attack. Still, Mussolini knew Great Britain was not willing to go to war over the issue.

On October 3, 1935, Italian troops and aircraft attacked Ethiopia. The League of Nations condemned Italy's actions, pledging to impose economic sanctions on Italy if it invaded Ethiopia. Under these sanctions, countries would not trade freely with Italy. But the league's measures had little effect. Over the next several months, Italian troops used poison gas and other

THE ROMAN EMPIRE

For about one thousand years, the Roman Empire ruled much of western Europe, along with parts of North Africa and the Middle East. Centered around the city of Rome, this empire is remembered as one of history's greatest civilizations. Its artistic, legal, and engineering achievements are still admired.

According to tradition, Rome was founded in 753 B.C. and was first under the control of a nearby kingdom. But Romans rebelled against their rulers around 500 B.C. They formed a republican government, a type of democracy. Then, in the 100s B.C., the republic was replaced by a dictatorship led by emperors called Caesars. One of the most famous of these powerful dictators was Julius Caesar.

During the next centuries, the Roman Empire used its military might to expand throughout Italy and into neighboring lands, until it stretched from southern England to modern Israel. Roman engineers built great roads and bridges to connect their conquered lands. They also built large government buildings and temples of stone and marble. Even long after the empire finally collapsed in the A.D. 400s, the ruins of these buildings remain throughout Europe. Rome's Latin language became the basis for Italian, Spanish, French, Portuguese, and Romanian.

Mussolini believed that his Italy was a new Roman Empire, and thought of himself as a new Caesar. He modeled the Fascist salute on the Roman salute and did many things to highlight historical connections between the two. Like the Caesars, Mussolini sponsored many building projects to commemorate his reign. As he grew older, his shaved head made him look more like some of the Caesars shown in ancient Roman art—and also helped hide the fact that he was losing his hair. He wore military-style uniforms and sometimes appeared on horseback. "I love Caesar," Mussolini once said of Julius Caesar. "He . . . combined the will of the warrior with the genius of a sage."

weapons to march through the country, capturing the capital in May 1936. Among the troops were Mussolini's sons Vittorio and Bruno, who served as bomber pilots. "Italy finally has its empire," Mussolini declared when his forces triumphed.

Italians celebrated wildly when victory was announced. Mussolini was more popular than ever. But the war and new territory had negative effects not seen at the time. They strained the Italian economy and encouraged corruption. And they may have made Mussolini think Italy was stronger than it really was.

IL DUCE'S SOFTER SIDE

In news stories and public appearances, Mussolini appeared to be a strong, unyielding leader. He had a warm personal side, however, and could be very loving toward his family. In June 1936, the dictator learned that his six-year-old daughter, Anna Maria, had polio. This devastating disease can cripple or kill its victims, and it had no cure. Mussolini spent a week at his sick daughter's bedside. Aides who saw him said his eyes were red with tears. But finally, Anna Maria started to recover. Her father, grateful, returned to work.

Mussolini prided himself on working hard. He told a newspaper reporter that he worked more than twelve hours a day, and read classic works and new novels as well as a stack of government papers every day. While this may have been an exaggeration, Mussolini did spend much of his time with his sleeves rolled up in his office, reviewing reports and personally directing the government.

THE SPANISH CIVIL WAR

Mussolini hoped that Fascism would spread across Europe, and in support of his movement, he sometimes helped foreign leaders who were Fascists or whose policies were similar to Fascism. For example, he sent nearly fifty thousand troops to Spain to help General Francisco Franco and members of Spain's Fascist Falange Party after they rebelled against the democratically elected government. Hitler also gave Franco some assistance.

Fascists fight government troops during the Spanish Civil War in 1936. Mussolini supported the Fascist rebellion, sending troops and weapons to aid its cause.

Spain's brutal civil war dragged on for three years. In the end, Franco's Fascists won. Franco established a dictatorship similar in many ways to Mussolini's. But the conflict cost Italy millions of dollars and pushed Mussolini and Hitler even closer. It also drew foreign criticism of Italian Fascists for their role in the devastation.

In September 1937, Mussolini traveled to Germany for the first time. Hitler and the other Nazis made every effort to impress him. Despite driving rain, eighty thousand people attended a speech Mussolini gave in German in the capital city of Berlin. Germany had grown much stronger in the three years since Hitler had risen to power. Realizing this, Mussolini adjusted his foreign policies. After appointing his son-in-law Ciano as foreign minister the year before, he had advised him that "it is in the interest of Italy that Austria remains independent." But he also said that he knew it would not remain independent forever— and that it would not be worth going to war with Hitler over it. In the end, he seemed to believe, there would be nothing Italy could do about Austria.

In December Italy formally left the League of Nations. No longer would Mussolini even pretend to go along with the organization. And in March 1938, when Hitler sent Mussolini a note saying that he had decided to march into Austria, Mussolini made no attempt to stop him. In return, Hitler sent a note of thanks. "Mussolini," he wrote, "I shall never forget this."

Self-proclaimed dictator and supreme commander of the military Benito Mussolini in 1938

Chapter FIVE

WORLD WAR

ON THE AFTERNOON OF SEPTEMBER 29, 1938, TEN men gathered in a room in Munich, Germany. The men included the leaders of France, Great Britain, Germany, and Italy. They had come to discuss whether or not the world would go to war.

Adolf Hitler had placed an army on the borders of Czechoslovakia (modern-day Czech Republic and Slovakia). Claiming that German-speaking citizens there were being discriminated against and attacked, he threatened to overrun the country. Great Britain and France had pledged to declare war if he did—but first they were making one last attempt at peace.

The men talked for hours. Though Mussolini had secretly assured Hitler that Italy would join Germany if

From left to right, front: *Neville Chamberlain (Britain), Édouard Daladier (France), Adolf Hitler (Germany), and Benito Mussolini. The European leaders met in Munich in 1938 to avoid war.*

there were total war, he was viewed by the others as a mediator. He also happened to be the only man in the room who spoke and understood all four languages.

At the end of the talks, the men agreed that Germany would occupy part of Czechoslovakia and there would be no war. It was a tremendous victory for Hitler. But the British were also happy with the agreement, believing that it would ensure peace for many years to come. The only unhappy country was Czechoslovakia, which lost a great deal of its territory to its much stronger neighbor without a fight—and without even being represented at the meeting.

When Mussolini returned home, his train was mobbed by Italians thrilled that he had prevented war. In the following weeks, world leaders also praised Mussolini's efforts. The British prime minister, Neville Chamberlain, suggested that their countries should work together more closely, and one of Hitler's ministers asked Mussolini to sign a military alliance. In both cases, Mussolini graciously but firmly said no.

THE EXPERIENCE OF ITALY'S JEWS

Anti-Semitism was a large aspect of Nazism. To Hitler and most of his followers, Jews were an inferior race. While Mussolini was prejudiced against Jews, the Fascists did not persecute the jewish population. "Anti-Semitism does not exist in Italy," Mussolini once declared. "Italians of Jewish birth have shown themselves good citizens and they fought bravely in [World War I]."

As his alliance with Hitler deepened, however, Mussolini had a document called the "Manifesto of Racial Scientists" prepared. It claimed that Italians—like Germans—were Aryans and therefore members of a master race, while Jews were not. Unlike Hitler, Mussolini did not order Jews to be rounded up, killed, or shipped to concentration camps and his followers did not systematically destroy their homes and businesses. But in 1938, he had passed laws banning Jews from marrying other Italians and barring them from a number of jobs, including teaching and scientific work. Non-Italian Jews were placed in internment camps, and discrimination against Jews increased. Meanwhile, in Germany, violence that would lead to the murder of millions of Jews in Europe also was increasing.

Experts have debated why Jews were not persecuted as a group in Italy as much as in Germany and elsewhere. One theory is that while Mussolini was personally prejudiced against Jews, he was not fanatical about hurting them or removing them from the country, as Hitler was. Perhaps this was because, on the whole, Italians were less anti-Semitic than people in countries such as Germany, where years of propaganda had called Jews enemies of the state. Experts also say that Pope Pius XI, who died in February 1939, initially helped convince Mussolini not to harm Jews.

POLAND AND THE WORLD WAR

The agreement to give Germany part of Czechoslovakia did not provide "peace in our time," as British prime minister Neville Chamberlain had promised in a famous speech. In fact, peace ended abruptly just a few months later when Hitler marched his armies into the rest of Czechoslovakia in March 1939. Chamberlain and other world leaders finally realized that Hitler would not stop until he conquered the entire world—or was conquered himself. Great Britain and France warned that they would declare war on Germany if Hitler attacked again.

While other countries were concerned about Hitler, Mussolini acted to expand his own empire. On April

A German tank division on parade in Warsaw, Poland, in 1939. The soldiers are receiving honors for their part in Nazi Germany's capture of Poland.

7, Italian troops invaded Albania, a small country across the sea from eastern Italy. They won a quick victory. The next month, Germany and Italy signed a treaty called the Pact of Steel, which made the two nations military allies. Mussolini now saw that war was inevitable. But he told Hitler that it should be delayed until at least 1942, while Italy prepared by building more battleships and other weapons.

Hitler was not about to wait. On September 1, 1939, German troops invaded Poland. Two days later, France and Great Britain declared war. But they could do little to help the Poles. By the end of the month, Poland had been conquered.

CIANO AND GERMANY

Italy remained neutral following the German invasion of Poland. The reasons were practical. The military was not prepared to fight, and Poland was thousands of miles away. The Italian army, navy, and air force had been stretched thin by the actions in Africa, Albania, and Spain. But Italy's biggest problem in waging war was that it couldn't produce the weapons and other materials needed to fight. Its armed forces were small and poorly equipped. Soldiers wore shoes made of cardboard instead of heavy boots. Many military and government leaders were aware of these shortcomings and urged Mussolini to remain neutral. "Il Duce is convinced of the need to remain neutral but he is not pleased about it," wrote Ciano.

Mussolini's son-in-law Ciano was now an important member of the government. Unlike Mussolini, he was not happy to be cooperating with Hitler, and he hated many of the high-ranking Germans he had to deal with. This difference in views caused much friction behind the scenes, especially with Germans who hated Ciano as much as he hated them. While Mussolini trusted his son-in-law and at times treated him as his own real son, in the end it was the dictator's word that was law. Italy drew ever closer to Germany.

Mussolini believed that Germany would easily defeat France and Great Britain. If Italy did not join the war, it would not share the glory when the war was over. Italy would also be in danger of being taken over by

Mussolini wearing formal military clothing. Joining forces with Nazi Germany during World War II, Mussolini hoped to dominate Europe.

Germany. Above all, he believed Italy was a great nation, and being a great nation meant going to war.

ALL-OUT WAR

In April 1940, Hitler invaded Denmark and Norway, quickly conquering these northern European countries. The next month, he launched an attack into Belgium and Holland and from there on to France. Within weeks, the French army was retreating madly to the south of their country. The British were on the run and fled across the English Channel.

On June 10, Mussolini pushed open the large glass doors on the balcony of his office in Rome. As the crowd in the street below chanted, "Duce! Duce!"

he declared war on Great Britain and France. "People of Italy, to arms, and show your tenacity, your courage, and your valor," he thundered.

France and Italy shared a small border at northwestern Italy. Italian troops prepared to attack across it. But before they could, the British bombed the Italian city of Turin in June. In retaliation, the Italian air force launched its own attacks on Malta, a Mediterranean island where the British had a base. Mussolini's son Bruno was one of the pilots, making his father proud with his bravery in combat.

As the Mediterranean war heated up, Bruno wasn't the only one of the Mussolini children in danger. Edda was serving as a volunteer nurse aboard a Red Cross hospital ship. When the British sunk the vessel, Edda and many other women had to abandon ship in the middle of the night. To the relief of her family, she was rescued and had not been seriously hurt in the attack.

After France surrendered to Germany on June 21, 1940, Hitler planned to bomb Great Britain into surrendering. Mussolini sent airplanes to help in the attacks on the British, but the small number of planes did not make much of an impact in the fighting. Hitler soon postponed his plans to invade Britain and began moving troops to the east of his country. He intended to invade the Soviet Union (a country made up of Russia and other republics), even though the countries had agreed a year earlier not to fight against each other. Hitler did not share his plans with Mussolini.

In August 1940, Mussolini took time from war planning and other government business to see his mistress Claretta Petacci in Rome. Claretta was very sick with peritonitis, an infection that sometimes follows operations. According to the official records, Claretta had had a miscarriage. However, most historians believe that she had actually had an abortion. Her husband was in Japan, and it is likely that the child was Mussolini's.

Greece

Hitler's armies won swift victories by combining airplanes with ground troops to make quick attacks—a new military tactic called blitzkrieg (lightning war). Admiring Germany's success, Mussolini decided to try the same approach against Greece. Italian armies crossed over the border from Albania into Greece on October 28, 1940. But the mountainous terrain was not well suited to blitzkrieg tactics. Even worse, the Italian army had neither the weapons nor the training to pull it off. The invasion was a disaster. The Greeks quickly counterattacked and drove the Italians back into Albania.

By winter the attack on Greece was going so badly for the Italians that Hitler decided to send his forces into the battle. As he saw it, because Germany and Italy were allies, the Greek victories made Germany look bad as well as the Italians. And the image that no one could defeat the Nazis—or their allies—was

German soldiers parachute into Greece in 1941. Mussolini's plans to easily capture the country were crushed by a determined Greek military. Hitler sent German troops into the country to save the failed invasion.

important, since it made it easier to intimidate and control other countries. The Italian defeat in Greece "struck a blow at the belief of our invincibility," Hitler said. So, in the spring of 1941, German armies attacked Greece and quickly took over the country.

The Italian troubles in Greece also lowered Hitler's personal opinion of Mussolini. "Mussolini was no longer the 'senior partner,'" wrote John Toland in his biography of Hitler. "In the Führer's eyes, he was branded with the unforgivable defect of failure." Other Germans put it even more bluntly. "We have the worst allies that could possibly be imagined," complained one of Hitler's top ministers, Joseph Goebbels.

GRIEF AND STRESS

Germany invaded the Soviet Union in June 1941. Mussolini congratulated Hitler on his victories as his armies swept toward the capital, Moscow. But privately, he told Claretta that he thought that the move would result in the Nazis losing the war. Mussolini believed that the Soviet Union was simply too large a country to attack successfully.

Meanwhile, the war was still going badly for the Italians. The British continued to bomb Italian cities. In North Africa, the British forced Italian troops in Libya and Egypt to retreat. Only when German troops joined the Italians in Africa did the fight go better there. Mussolini tried to find ways to improve Italy's fighting and increase war production. But British bombing and Italy's disorganized industries made accomplishing these tasks difficult.

Some biographers and historians believe that the Italian people began turning away from Mussolini during 1941, upset by the Italian army's failures in Greece and northern Africa. A report by a U.S. diplomat declared, "The people feel that against their own better judgment they were led by Il Duce and the party leaders to believe that Fascism had forged them into a race of warriors; now they knew they never were nor will be."

The war took a personal toll on Mussolini, as well. In August 1941, his son Bruno died in an airplane accident. Mussolini's grief at this loss, along with the stress of the war, added to his continuing stomach problems. He couldn't sleep at night because of terrible cramps. During the day, he often had to stop speaking and press his fist against his stomach to try to relieve the pain.

TURNING TIDE

In December 1941, Japan attacked U.S. naval base at Pearl Harbor, Hawaii. The United States declared war, joining France and Great Britain as part of the Allied powers. In turn, Germany and then Italy both declared war on the United States. The war now involved all of the major powers of the world, and every continent except for South America and Antarctica.

As the year ended, Germany seemed to be on the verge of great triumph. Despite Italy's troubles, German armies occupied most of Europe and were close to the capital of the Soviet Union. But in 1942, things slowly began to change. The Soviet Union's vast area and horrible winter weather presented a unique challenge to German troops. In addition, the United State's entry into the war pitted Germany and its allies against the world's largest industrial power. While it took the United States many months to train and equip a large army, its size and wealth dwarfed what Germany, Italy, and Japan could muster, even together.

In November of 1942, the United States joined British forces in North Africa with landings in Morocco and Algeria. The move caught both Hitler

German and Italian prisoners of war on the march in North Africa in 1942. These soldiers fell to superior Allied forces there, frustrating both Mussolini and Hitler.

and Mussolini by surprise. The Germans rushed airplanes to North Africa but were bogged down in the Soviet Union and could not spare a large army to reinforce their troops.

In late 1942, Mussolini sent Ciano to Germany for a series of meetings with Hitler that Mussolini was too ill to attend himself. Ciano tried to convince Hitler to retreat from the Soviet Union so German troops could help fend off the Allies in North Africa. Hitler refused. As U.S. and British forces chased German and Italian armies from northern Africa, the skies above Italy filled with U.S. bombers. Italian cities were attacked relentlessly. As the situation grew still bleaker, Communists began to stir up opposition to the government in Italy's northern industrial cities. Mussolini, entangled in other challenges, was unable to stop them.

With public opinion turning against il Duce, some people began discussing possible ways of removing Mussolini from power. The king—who still technically played a role in the government—had always gone along with whatever Mussolini wanted. As his advisers urged him to replace the prime minister, he refused to go against the constitution, which only allowed him to remove Mussolini if the Chamber of Deputies voted for a change. Since Mussolini controlled the Chamber, this was extremely unlikely. But discontent with the war and with Mussolini continued to brew.

Meanwhile, Hitler, worried that U.S. and British ground forces would invade Italy, offered to send five divisions of German troops to Mussolini in the spring of 1942. Mussolini refused, probably realizing that once large numbers of German troops came into the country, he would lose control. He did, however, accept some small German units and asked for more aircraft. He knew that the situation was getting desperate. He heard from his own daughter Edda, who sent a telegram in May from Palermo, a city in Italy's large southern island of Sicily. "Civilians feel themselves abandoned," she told her father. "Never have I seen such suffering and pain."

INVASION

On July 9, 1943, British and U.S. forces landed in Sicily. Italian troops were forced to retreat. A few days later, a massive U.S. bombing attack on Rome sent 150,000 Italians fleeing the city.

The Fascist Grand Council met on the night of July 24 to discuss the dire situation. Mussolini's wife Rachele advised him to arrest everyone at the meeting. He didn't take her advice. Instead, he seems to have thought he could persuade the council that things would get better. He gave a speech at the meeting admitting that the war had gone badly. But he pointed out that Italy had also lost territory during World War I, before finally rebounding to win.

Allied forces come ashore during the invasion of Sicily in 1943. The men received their "go" orders on July 9 of that year.

With the invasion of Sicily and the advance of the Allies through Italy, the Grand Council, above, lost faith in Mussolini's ability to command. They stripped him of his military authority.

The Grand Council wasn't swayed. After hours of debate, the council voted in favor of taking Mussolini's military authority away and giving it to the king.

"You have brought this regime into crisis!" Mussolini declared. "This meeting is adjourned." When the members of the council rose to give him the Fascist salute, he told them not to bother.

"I know very well where these matters will end," said Mussolini before going home. It was just four days before his sixtieth birthday.

ARREST

The next day, Mussolini went to work as usual. At 5 P.M., he left his offices and went to visit the king. As

Mussolini told King Victor Emmanuel III about the military situation and the vote by the Fascist Grand Council, the king interrupted him. "My dear Duce, it can't go on any longer. Italy is in pieces," said the king. "You are the most hated man in Italy."

"I am perfectly aware that people hate me," replied Mussolini. With that, the king escorted him to the door. Outside, an ambulance had pulled up and blocked his car. A police officer stopped him.

"His majesty has charged me with the protection of your person," he said.

"That's not necessary," said Mussolini.

"No, you must come in my car," said the policeman, leading him to the ambulance.

The supreme leader of Italy was under arrest.

THE CONDEMNED MAN

A much less powerful Mussolini, center, appeared on the cover of Picture Post magazine in 1943. Labeled "the condemned man" by the publication, Mussolini was certainly detested by many in Italy following his wartime failures.

Chapter **SIX**

THE BITTER END

THE VIEW FROM GRAN SASSO D'ITALIA—THE HIGHEST
mountain in the Italian Apennines—was breathtaking.
At the peak was an old hotel, a former ski lodge that
had been a spectacular vacation spot before the war.
Now it had only one guest—Benito Mussolini. He
spent his days playing cards with the policemen who
guarded him. When the early September nights turned
chilly, he talked of maybe going skiing.

Mussolini had been moved around quite a bit since
his arrest a little more than a month before. He had not
been mistreated, though he was clearly a prisoner. Most
Italians were happy to have him out of power. More
than a few wanted to take revenge on him for the terrible situation Italy now found itself in. While he claimed

to want to retire to his home in Forlì, government officials there insisted he would be lynched if he did return,
and he remained under house arrest.

When the new Italian government tried to surrender to
the Allies, German troops had marched into Italy to pre-
vent it from being seized by the enemy. The situation
was extremely chaotic, with both Italian and German
forces in charge. The Allies continued to fight both.

Hitler, acting out of friendship for Mussolini,
ordered German troops to take Mussolini's family out
of Italy and keep them safe. But his soldiers had not
found Mussolini.

Suddenly, on the afternoon of September 12, a formation of large gliders appeared over the mountain resort.
As the police watched, the silent aircraft landed and specially trained German troops jumped out. The leader of
the force, Otto Skorzeny, raced up the steps of the hotel
and found Mussolini waiting with two of his guards.

Skorzeny told Mussolini that Hitler had sent him to
rescue him.

"I knew all along that the Führer would give me this
proof of his friendship," replied Mussolini.

Mussolini was whisked to Germany. After medical
treatment for his old ailments, he met with Hitler. With
German troops now controlling a little more than half
the country and the Allies holding the rest, Hitler per-
suaded Mussolini to return to Italy and head the government. Mussolini reluctantly agreed, taking over the
German-controlled region of northern Italy—the Repub-

Sprung from Italian house arrest, Mussolini, front, *is whisked away to Germany by a Nazi commando unit on September 12, 1943.*

lic of Salò. But now he was only a puppet of Hitler. He wasn't even allowed in Rome by the German military commanders. At the same time, about seven hundred thousand Italian troops were captured and taken into Germany. They were considered traitors by the Nazis because the government had tried to surrender.

A PUPPET RULER

Rachele, the Mussolinis' daughter Anna Maria, and several grandchildren joined Mussolini in a villa on northern Italy's Lake Garda. Things were very different for il Duce now. With no real power, he had to do whatever Hitler and German military commanders wanted. Among their demands was that he put his son-in-law Ciano and a few others on trial for treason and have them shot. The Germans had found a diary in which Ciano had recorded his thoughts about them, and they wanted revenge. Edda pleaded desperately with Mussolini to save Ciano's life. But Mussolini had little choice. "If you will not deal with him," Hitler had warned, "we shall hang him here in Germany." And so Mussolini allowed his daughter's husband to be killed, even though he knew the charges were false. Edda never forgave her father, considering him Ciano's murderer.

Mussolini's return under German control opened the way for Hitler to target Italian Jews. Some brave Italians tried to protect Jews from German persecution, but between 7,000 and 10,300 Jews were eventually deported from Italy. Most were killed. As many as 9,000 died.

By this time, the Italian army had disintegrated. The war now was mostly fought by Germans, as the U.S. and British armies slowly moved up the peninsula toward Rome. In the meantime, the Italian people suffered. Food and jobs were scarce. Citizens lived in constant fear of bombings. Groups of Italian partisans —citizens who fought the Germans with whatever weapons they could find—attacked German bases and supplies. The Germans reacted brutally, killing innocent civilians as well as the partisans. These partisans were often Socialist and Communist opponents of Mussolini who, in many cases, had worked with other anti-Fascists throughout Mussolini's reign. When the partisans tried to kill Mussolini with a bomb in March 1944, the Germans responded by rounding up 335 men and machine-gunning them outside Rome.

"GREAT DESTINY"

In July 1944, Mussolini went to Germany. He met with Hitler just hours after a bomb planted by a German officer nearly killed the führer. Hitler even showed him where the bomb had gone off. But while Mussolini sensed that the end was near, Hitler did not. "I am more than ever sure that the great destiny I serve will transcend its present perils," Hitler told the Italian leader.

By the winter of 1944, U.S. and British troops controlled France. Soviet troops were marching toward Berlin. Only a small part of northern Italy around the Po River Valley, including Milan, remained in German hands. Mussolini made a defiant speech in Milan on December 16. "We will defend the Po valley with our nails and our teeth," he declared. That day he sounded like il Duce of old, but his glory days were long gone. As spring arrived, the Allies renewed their attacks. On April 21, 1945, they reached Bologna. With enemy troops closing in, many partisans and Italian civilians began killing Fascists and Germans. Although the United States and Great Britain had promised to punish Fascists, the partisans did not trust them to deliver justice.

At the same time, U.S. troops and specially trained members of the U.S. Office of Strategic Services (an undercover military intelligence group) began hunting for Mussolini. They wanted to capture him alive and put him on trial for crimes during the war. And they needed to reach him before the partisans—who wanted him dead—found him.

ESCAPE—AND CAPTURE

In late April 1945, the German army in Italy collapsed. Not sure what to do, Mussolini at first thought he could make a final stand against his enemies. Separated from Rachele, he sent her a letter telling her that she should flee with the children. "We may never see each other again," he wrote. But Mussolini quickly realized resistance was impossible. Hoping to escape to Switzerland, he joined a group of German soldiers.

Before Mussolini reached the Swiss border, however, partisans stopped his convoy of cars and trucks. After a brief gun battle, the partisans said they were not interested in the Germans—only the Fascists. Finally, the Germans agreed to surrender all the Italians. Mussolini hid in the back of one of the trucks, wearing a German coat and helmet. When the truck was searched, the Germans said he was one of their soldiers. But the partisans weren't fooled. They seized Mussolini. Claretta, who'd been warned by Mussolini to escape but had decided to follow him, was captured in a nearby car.

The partisans acted quickly and brutally. The very next day—April 28, 1945—Mussolini and Claretta were shot. Their bodies were driven overnight to Milan, where they were flung down in a square that had been used by Germans to execute fifteen partisans the year before.

A mob beat, spit at, and even defecated on the bodies of Mussolini and his mistress. The battered corpses were hung upside down by their feet for all to see. Mussolini's reign of Fascism had finally come to an end.

A public poster of Mussolini in Milan, Italy, is riddled with bullet holes in 1945. Taking out their anger over years of Fascist rule, many Italians defaced images of the dictator. Following his execution in April 1945, a mob also took out their anger on Mussolini's body.

WORLD WAR II

orld War II ended in Europe in May 1945 after Hitler committed suicide, and Allied armies took over Germany. In August Japan surrendered after the United States dropped atomic bombs on the cities of Hiroshima and Nagasaki.

This war was the most destructive conflict ever fought. Nearly every major country in the world took part in the war. The number of deaths caused by the war cannot be counted precisely, but estimates have ranged from 40 to 50 million and up—the majority of whom were civilians. Italy lost approximately 410,000 of its people. The war left Europe and much of Asia, including Japan, China, and the Philippines, in ruins.

MUSSOLINI'S LEGACY

World War II continued in Europe for only a few days more. In Asia fighting continued until the United States dropped two atomic bombs on Japan in August 1945.

The collapse of Germany and its allies freed millions of people from oppression. But the war left much of the world in ruins. It took many years for Italy to recover from the devastation.

Fascists were out of power in Italy, but the ideas of Fascism did not disappear entirely with Mussolini's death. An Italian political party called Movimento

Sociale Italiano (Italian Social Movement) kept alive many of Mussolini's ideas after the war. Although it did not play an important role in the government, it received about 8 percent of the vote in elections following the war. In Spain Franco continued to govern until his death in 1975.

In modern times, no government openly calls itself Fascist or claims to use Mussolini as a model. Still, many dictators use tactics similar to Mussolini's. Some governments, such as Syria's, allow only one political party to operate. And leaders still employ violence to intimidate their critics and use fear to win support from average people.

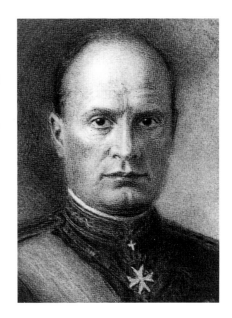

This portrait of Mussolini was created in 1925, when Il Duce was strong and his Fascist vision for Italy and Europe was clear. Although some of the things he did helped Italy, Mussolini's reign was marred by violence, unprovoked war, and the abuse of power.

Looking at Italy's failures in World War II, some modern observers view Mussolini almost as a joke. Citing the corruption during his reign, they ignore his successes with public works projects and the Italian economy. They forget that his methods appealed to a great number of people in a democracy under strain.

Above all, Mussolini reminds us that even in a democracy, bad leaders can come to power. They may seem intelligent and even charming—but if they are allowed to have unlimited power, the results will be catastrophic. And perhaps that lesson is his true legacy. "I think . . . that there will not be a second 'Duce,'" said Mussolini, well before his time in power ended. "If he appeared upon the scene, Italy would not put up with him."

SOURCES

7   R. J. B. Bosworth, *Mussolini* (New York: Oxford
    University Press, 2002), 52.

13 Ibid., 43.

14 Benito Mussolini, *My Autobiography* (New York: Charles
    Scribner's Sons, 1928), 4.

17 Emil Ludwig, *Talks with Mussolini* (Boston: Little, Brown,
    and Company, 1933), 49.

18 Bosworth, 66.

19 Mussolini, *My Autobiography*, 15.

19 Ibid., 15.

19 Bosworth, 71.

20 Rachele Mussolini and Albert Zarca, *Mussolini: An
    Intimate Biography by His Widow, Rachele Mussolini, as
    Told to Albert Zarca* (New York: Pocket Books, 1977), 21.

21 Ibid.

21 Ibid.

22 Bosworth, 85.

23 Ibid.

25 Laura Fermi, *Mussolini* (Chicago: University of Chicago
    Press, 1961), 56.

25 Bosworth, 86.

26 Ibid., 95.

28 Ibid., 104.

30 Mussolini, *My Autobiography*, 43.

30 Ibid., 43–45.

30 Ludwig, 200.

30 Ibid.

31 Fermi, 135.

31 Mussolini, *My Autobiography*, 55.

32 Ibid., 70.

34 Ibid., 68.

35 Ibid., 99.

35 Ibid.

37 Bosworth, 152.

x38 Edda Mussolini Ciano and Albert Zarca, *My Truth* (New
    York: Morrow, 1977), 34.

38 Ibid., 44.
44 Ludwig, 93.
49 Ian Kershaw, *Hitler, 1889–1936: Hubris* (New York: W. W. Norton, 1999), 131.
50 Bosworth, 176.
52 Mussolini, *My Autobiography*, 233.
52 Ibid.
54 Fermi, 249.
55 Bosworth, 225.
57 Jasper Ridley, *Mussolini* (New York: St. Martin's Press, 1998), 208.
57 William Bolitho, *Italy under Mussolini* (New York: Macmillan, 1926), 2.
57 Ibid., 3.
60 Bosworth, 238
66 Ibid., 280.
67 Ibid., 296.
67 Ludwig, 222.
67 Ibid.
70 Ibid., 62.
71 Bosworth, 309.
73 Ibid., 330.
73 Ibid., 331.
77 Ludwig, 70.
77 Ibid.
80 Bosworth, 356.
82 Ridley, 314.
84 John Toland, *Adolf Hitler* (New York: Ballantine Books, 1981), 651.
84 Ibid., 654.
84 Bosworth, 379.
85 Edwin P. Hoyt, *Mussolini's Empire: The Rise and Fall of the Fascist Vision* (New York: John Wiley & Sons, 1994), 202.
88 Bosworth, 398–399.
88 Ibid.
90 Benito Mussolini, *The Fall of Mussolini: His Own Story* (New York: Farrar, Straus, 1948), 66.
90 Bosworth, 401.
91 Mussolini, *The Fall of Mussolini*, 71.

91   Ibid., 72.
91   Hoyt, 229.
94   Mussolini, *The Fall of Mussolini,* 139.
96   Hoyt, 258.
96   Nora Levin, *The Holocaust: The Destruction of European Jewry, 1933–1945* (New York: Schoken Books, 1973), 468.
97   Hoyt, 267.
97   Ibid., 272.
98   Mussolini and Zarca, *Mussolini,* 281.
102   Ludwig, 133.

SELECTED BIBLIOGRAPHY

Bolitho, William. *Italy Under Mussolini*. New York: Macmillan, 1926.

Bosworth, R. J. B. *Mussolini*. New York: Oxford University Press, 2002.

Ciano, Edda Mussolini, and Albert Zarca. *My Truth*. New York: Morrow, 1977.

Deakin, F. W. *The Brutal Friendship: Mussolini, Hitler and the Fall of Italian Fascism*. Rev. ed. London: Phoenix Press, 2000.

Duggan, Christopher. *A Concise History of Italy*. New York: Cambridge University Press, 1994.

Fermi, Laura. *Mussolini*. Chicago: University of Chicago Press, 1961.

Hoyt, Edwin P. *Backwater War: The Allied Campaign in Italy, 1943–1945*. Westport, CT: Praeger, 2002.

———. *Mussolini's Empire: The Rise and Fall of the Fascist Vision*. New York: John Wiley & Sons, 1994.

Kershaw, Ian. *Hitler, 1889–1936: Hubris*. New York: W. W. Norton, 1999.

"Italian Life Under Fascism: Selections from the Fry Collection." *The Fry Collection of Italian Fascist Publications, University of Wisconsin-Madison*. 1998. http://www.library.wisc.edu/libraries/dpf/Fascism (April 26, 2005).

Levin, Nora. *The Holocaust: The Destruction of European Jewry, 1933–1945*. New York: Schoken Books, 1973.

Mack Smith, Denis. *Mussolini*. New York: Alfred A. Knopf, 1982.

Morgan, Ted. *FDR: A Biography*. New York: Simon and Schuster, 1985.

Mussolini, Benito. *The Fall of Mussolini: His Own Story*. New York: Farrar, Straus, 1948.

———. *My Autobiography*. New York: Charles Scribner's Sons, 1928.

Mussolini, Rachele, and Albert Zarca. *Mussolini: An Intimate Biography by His Widow, Rachele Mussolini, as Told to Albert Zarca*. New York: Pocket Books, 1977.

Ridley, Jasper. *Mussolini*. New York: St. Martin's Press, 1998.

Toland, John. *Adolf Hitler*. New York: Ballantine Books, 1981.

FURTHER READING AND WEBSITES

BOOKS

Behnke, Alison. *Italy in Pictures*. Minneapolis: Lerner Publications Company, 2003.

Black, Wallace B., and Jean F. Blashfield. *Invasion of Italy*. New York: Maxwell Macmillan International, 1992.

Downing. David. *Fascism*. Chicago: Heinemann Library, 2003.

Feldman, Ruth Tenzer. *World War I*. Minneapolis: Lerner Publications Company, 2004.

Goldstein, Margaret J. *World War II: Europe*. Minneapolis: Lerner Publications Company, 2004.

Roberts, Jeremy. *Adolf Hitler: A Study in Hate*. New York: Rosen, 2001.

Zuehlke, Jeffrey. *Germany in Pictures*. Minneapolis: Lerner Publications Company, 2003.

WEBSITES

"Benito Mussolini." *Commando Supremo: Italy at War.* http://www.comandosupremo.com/Mussolini.html. This site provides extensive information on Mussolini, Fascist Italy, and the history of Italy's role in World War II.

"Italian Life under Fascism: Selections from the Fry Collection." *Department of Special Collections: University of Wisconsin-Madison.* http://www.library.wisc.edu/libraries/dpf/Fascism. Learn more about the history of Fascism, including information about Mussolini and Fascist Italy, as well as view documents and art from the period.

"Benito Mussolini, Dictator of Italy, Declares War on Britain and France." *History Channel: Speeches.* http://www. historychannel.com/speeches/archive/speech_378.html. Read and listen to Mussolini's declaration of war on Britain and France from June 10, 1940.

INDEX

OTHER TITLES FROM LERNER AND A&E®:

Ariel Sharon
Arnold Schwarzenegger
Arthur Ashe
The Beatles
Benjamin Franklin
Bill Gates
Bruce Lee
Carl Sagan
Chief Crazy Horse
Christopher Reeve
Colin Powell
Daring Pirate Women
Edgar Allan Poe
Eleanor Roosevelt
Fidel Castro
Frank Gehry
George Lucas
George W. Bush
Gloria Estefan
Hillary Rodham Clinton
Jack London
Jacques Cousteau
Jane Austen
Jesse Owens
Jesse Ventura
Jimi Hendrix
J. K. Rowling
John Glenn
Latin Sensations

Legends of Dracula
Legends of Santa Claus
Louisa May Alcott
Madeleine Albright
Malcolm X
Mark Twain
Maya Angelou
Mohandas Gandhi
Mother Teresa
Nelson Mandela
Oprah Winfrey
Osama bin Laden
Pope John Paul II
Princess Diana
Queen Cleopatra
Queen Elizabeth I
Queen Latifah
Rosie O'Donnell
Saddam Hussein
Saint Joan of Arc
Thurgood Marshall
Tiger Woods
Tony Blair
Vladimir Putin
William Shakespeare
Wilma Rudolph
Women in Space
Women of the Wild West
Yasser Arafat

ABOUT THE AUTHOR:

Jeremy Roberts has written a number of biographies of World War II and Holocaust figures, including Adolf Hitler, Franklin Roosevelt, and Joseph Goebbels.

PHOTO ACKNOWLEDGMENTS

The images in this book are used with the permission of: courtesy of the Library of Congress, pp. 2 (LC-USZ62-118861), 6 (LC-USZ62-118861), 87 (LC-USZ62-070812), 101 (LC-USZ61-678; © Rischgitz/Getty Images, p. 8; © CORBIS, pp. 11, 13, 90, 95; © Bettmann/CORBIS, pp. 14, 24, 40, 81; © Pix Inc./Time Life Pictures/Getty Images, pp. 18, 31, 42, 44; © Hulton Archive/Getty Images, pp. 23, 56, 64, 89, 99; courtesy of the United States Holocaust Memorial Museum (USHMM), p. 36; © Topical Press Agency /Hulton Archives/Getty Images, p. 54; © Keystone/Getty Images, pp. 69, 74, 84; © STF/AFP/Getty Images, p. 72; © Independent Picture Service (IPS), p. 76; United States Holocaust Memorial Museum (USHMM), courtesy of Richard A. Ruppert, p. 79; © IPC Magazines/Hulton Archives/Getty Images, p. 92.

Front cover: courtesy of the Library of Congress (LC-USZ62-79593). Back cover: courtesy of the Library of Congress (LC-USZ62-73360).

WEBSITES

Website addresses in this book were valid at the time of printing. However, because of the nature of the Internet, some addresses may have changed or sites may have closed since publication. While the author and Publisher regret any inconvenience this may cause readers, no responsibility for any such changes can be accepted by the author or Publisher.